YOUR KNOWLEDGE HAS VALUE

Ralph Myers

Goal-Oriented Humanitarianism and Colombia's Civil War

Developing Strategies Towards Ending and Preventing Civil Conflict in Colombia

GRIN Verlag

Bibliografische Information der Deutschen Nationalbibliothek:

Die Deutsche Bibliothek verzeichnet diese Publikation in der Deutschen National-
bibliografie; detaillierte bibliografische Daten sind im Internet über http://dnb.d-
nb.de/ abrufbar.

Imprint:

Copyright © 2011 GRIN Verlag GmbH
Druck und Bindung: Books on Demand GmbH, Norderstedt Germany
ISBN: 978-3-656-44237-0

This book at GRIN:

http://www.grin.com/en/e-book/215493/goal-oriented-humanitarianism-and-
colombia-s-civil-war

GRIN - Your knowledge has value

Der GRIN Verlag publiziert seit 1998 wissenschaftliche Arbeiten von Studenten, Hochschullehrern und anderen Akademikern als eBook und gedrucktes Buch. Die Verlagswebsite www.grin.com ist die ideale Plattform zur Veröffentlichung von Hausarbeiten, Abschlussarbeiten, wissenschaftlichen Aufsätzen, Dissertationen und Fachbüchern.

Visit us on the internet:

http://www.grin.com/

http://www.facebook.com/grincom

http://www.twitter.com/grin_com

Goal-Oriented Humanitarianism and Colombia's Civil War

Developing Strategies Towards Ending, and Preventing, Civil Conflict in Colombia

Ralph Myers

06/01/2011

<u>Assignment Submitted in partial fulfilment of the Geopolitics Module to Module</u>

<u>1st semester of Masters in Humanitarian Action Programme 2010-2011, UCD College of</u>

<u>Life Science</u>

Introduction

Humanitarian organisations have always realised that their activities have geopolitical consequences and that they are inseparably linked to the political world. However, since the 1990's, humanitarian organisations have increasingly begun acting upon this awareness (Barnett 2005: 724). The watershed for this transformation from minimalist assistance provision to goal-oriented humanitarianism, was the decision of numerous humanitarian organisations to pull out of Rwandan refugee camps in Zaire. The argument was that these camps were being used by Hutu *genocidaires* to launch violent attacks on Rwandan soil (Fox 2001: 279-280, 285-288). The shift to goal-oriented humanitarianism[1] has resulted in "a broad trend towards an increased use of humanitarian assistance as part of a more comprehensive strategy to transform conflicts and decrease the violence" (Uvin 1999: 8). This paper identifies three strategies in which humanitarian organisations can address the root causes of conflict in the 21st century: disaster risk reduction (DRR), the decision to withhold or focus aid in specific areas, and speaking out. DRR is usually executed pre-conflict or during periods of reduced violence, whilst the latter two are relevant at the time of conflict.

The main aims of this paper are to examine the phenomenon of civil war from a geopolitical perspective and identifying variables linked to both the onset and duration of civil conflict, specifically in the case of Colombia. The conclusions drawn here should assist humanitarian organisations present in Colombia, in identifying ways to implement the above three mentioned strategies so as to address the root causes of civil war. Halting conflict and preventing future reoccurrence of civil war is in the direct interest of Colombian based humanitarian organisations, as many of them work with internally displaced persons (IDPs). As Ibáñez and Vélez (2008: 672) note: "Violence and security perceptions are the major determinants of displacement and are, therefore, the key instruments in preventing displacement."

This paper begins by examining Colombia's ongoing conflict from a historical perspective, identifying the main actors as well as the impacts of violence on its civilian population. The chapter thereafter looks at how non-governmental organisations (NGOs) deal with the humanitarian consequences of Colombia's civil war, specifically in relation to IDPs. An impoverishment risk and livelihood reconstruction (IRLR) model is used to identify which main areas humanitarian organisations do, and should, focus on, with regards to providing (developmental) relief.

---

[1] Also referred to as developmental relief

The fourth and fifth paragraphs identify relevant independent variables for civil war onset and duration, drawing from the two main schools of thought within the civil conflict literature. These two schools of thought are divided amongst so called 'grievance' and 'greed' advocates. The first focus on grievances arising from a number of possible sources of relative deprivation amongst ethnic groups. Grievance theories are mainly grounded in qualitative analyses. This paper focuses on the works of Gurr (1970, 2000) and Stewart (2002, 2010). Greed theories of civil war focus more on individual incentives of rebels, most notably economic gain. These theories are often presented with the use of a quantitative model, the most famous and influential being that of Collier and Hoeffler (2004). This paper focuses both on civil war onset, as well as duration. This is done for two reasons: first, focusing on civil war onset is useful for the implementation of future DRR strategies, second some codifications for civil war used in this paper, e.g. Gleditsch' (2004) revision of the Correlates of War (COW) project,[2] have recorded relatively recent periods of 'peace'. This paper also argues that although studies of civil war onset and duration are not the same, the similarities between the two are so great that they are analysed together here.

By examining both theories of civil war, we should be able to identify which theory best explains the phenomenon of civil conflict in Colombia, as well as which independent variables causing civil war are addressed in the IRLR model. Humanitarian organisations can then employ this knowledge directly with regards to providing relief and indirectly whilst implementing DRR and doing advocacy work.

This paper puts forward the following hypotheses:

H1:   *Greed theories of civil war better explain the Colombian conflict*

H2:   *Goal-oriented humanitarian organisations should focus on those components found in Muggah's (2000) adapted IRLR model, which correspond to independent variables identified in greed theories of civil war relevant to the Colombian context, in order to effectively implement strategies of relief provision, in addition to DRR and advocating*

---

[2] There is now generally agreed definition for civil war, most qualitative studies mention a period of over forty years of armed conflict (UNHCR 2010: 3), whilst the quantitative studies used in this paper denote the year 1984 for the start of civil war in Colombia. (Collier & Hoeffler 2004: 566, Collier et al. 2009: 7)

Colombia's Civil War: Historical Context, Actors and Impacts

In order to adequately understand the current day internal conflict in all its aspects, one must first understand the historical context of Colombia's civil war,[3] its actors and its resulting impacts on the civilian population.

Colombia, South America's fourth largest country, won its independence from Spain in 1819. Thirty years after independence, the Conservative and Liberal parties were founded. Disputes between followers of both parties led to small episodes of violence up until 1899. Ideological differences existed between landowners from Spanish descent who held both economic and political power. The peasants working on the farms of these landowners traditionally adopted ideologies similar to their employer. (BBC 2010a, BBC 2010b, Rogers 2002: 4-5)

From 1899 on there were a number of episodes of conflict, culminating in a particularly brutal period referred to as 'la Violencia', starting in 1948. (Rogers 2002: 5). The time period of 'la Violencia' sowed the seeds for the formation of the two most notable left-wing guerrilla movements: Ejercito de Liberacion Nacional (ELN) and the Fuerzas Armadas Revolucionarias de Colombia (FARC) (Minear 2006: 9, Rogers 2002: 5).

The 1970's saw a dramatic rise in poverty, specifically in rural areas, in addition to the continued inability of the Colombian government to address inequalities of land distribution. At the same time the global demand for cocaine was on the rise, and many peasants moved into FARC controlled territories to grow coca plants (Leech 2002: 17). The seventies also saw the formation of right-wing paramilitary groups. They were formed as a result of land owners and businessmen wanting to defend their capital from guerrilla movements and advance their own interests in the absence of a strong state structure. These paramilitary movements were able to grow significantly through their involvement with the drug trade as well as being given free reign by the military. (LeoGrande & Sharpe 2000: 5). With regards to recruitment, the paramilitaries used similar tactics to those of left-wing guerrilla groups and recruited from the marginalised peasant classes. A number of these small paramilitary groups came together in 1997 to form the Autodefensas Unidas de Colombia (AUC). Together with the left-wing guerrillas and the government they form the three principle (armed) actors to the Colombian conflict (Minear 2006: 10).

In the 1980's, the FARC attempted to form its own political party – the Unión Patriótica – which was heavily targeted by right-wing paramilitaries, resulting in the deaths of

---

[3] The Colombian government does not view it's civil conflict as a war, however most of the quantitative studies referred to in this paper do codify it as such.

over 3,000 party members, making the party unviable. During this time, the global demand for illicit narcotics such as cocaine and heroine allowed for an increase in financing of both right and left-wing armed groups (LeoGrande & Sharpe 2000: 4).

The end of the Cold War and the 9/11 terror attacks have also had a large influence on the civil war in Colombia, specifically with regards to United States' political, military and financial assistance to the government of Colombia. With the winding down of the Cold War, US priorities shifted from combating Marxism to combating the perceived threat of narcotics trafficking. The global war on terror, which the US launched in the wake of 9/11, once again shifted the paradigm from combating 'narco-trafficking' to the fight against 'narco-terrorism'. The culmination of US assistance is the realisation of 'Plan Colombia', which in theory is supposed to address a wide array of issues in Colombia but in practice weighs heavily on security (Minear 2006 12-14).

According to the International Committee of the Red Cross (ICRC), the civil conflict in Colombia, as of 2009, has led to the displacement of up to 4,360,000 people (Carillo 2009: 527). This has resulted in a major humanitarian and human rights crisis. Many of the IDPs and refugees are without identity papers or formal titles to their land. Colombia's Indigenous and Afro-Colombian populations are being particularly affected. The human rights situation is deemed to be particularly grave and includes violations of both political and civil rights (Minear 2006: 1-2, 18-19). It is undeniable that Colombia's civil conflict and responses to it such as aerial spraying are primarily responsible for the humanitarian crisis of displacement. Ibáñez and Vélez (2008:661-662).

Using the IRLR Model in Addressing both Humanitarian Needs and the Causes of Conflict.

As the last section of the previous paragraph shows: there are many different aspects of, and implications arising from, displacement. Cernea's (1997) impoverishment risk and livelihood reconstruction (IRLR) model, and more specifically Muggah's (2000) adaptation to conflict scenarios, provides a good starting point from which to identify areas different humanitarian organisations currently do, and should, focus on in relation to displacement. Both models contain components that resemble, and build on, the *Guiding Principles on Internal Displacement* (Deng 1998). Muggah's conflict induced displacement model contains 11 variables relating to impoverishment risk resulting from displacement. These include; "landlessness, joblessness, homelessness, economic marginalisation, increased morbidity, food insecurity, loss of access to common property, (…) social disintegration, (…) limited access to education, declines in political participation and the increased risk of political and criminal violence" (Muggah 2000: 199-200). These impoverishment risks are coupled to 11 components designed to reverse such risks, such as; land based re-establishment, re-employment, social integration, reformation of political activity and protection, etc. (Muggah 2000: 200).

The following sections will identify independent variables related to the two dominant theories of civil war, namely greed and grievance. Both theories identify different independent variables which have a significant impact with regards to the onset and duration of civil war. A number of these variables correlate to areas identified in Muggah's adapted IRLR model as important for humanitarian organisations to focus on whilst providing relief. In effect, this means that certain impoverishments resulting from conflict induced displacement, are in fact fuelling the civil war itself, creating a downward spiral of violence. Identifying which of these correlations are most relevant to the current situation in Colombia, will greatly aid developmental relief organisations in efficiently and purposefully implementing strategies of aid allocation. In addition, the findings of this paper can support advocacy work and future disaster risk reduction.

Grievance Theories of Civil War

The greed versus grievance dichotomy was first distinguished by Paul Collier (2000), as a result of a new trend within civil war studies. These scholars, whilst employing econometric models of analysis, focused more on individual rebel behaviour referred to as "greed". Grievance theories by contrast, focus more on the group level and are usually qualitative. Rebellion can be viewed as a means to redress certain grievances, which include social (e.g. access to education), political (e.g. political participation), economic (e.g. ownership of assets) or cultural (e.g. recognition of language) inequalities (Stewart 2010: 7). Most recently this has translated into the study of so called 'horizontal inequalities' by scholars such as Frances Stewart (2002, 2010) and Gudrun Østby (2008). Their work focuses on relative deprivation amongst groups with a strong identity, usually along ethno-linguistic and/or religious lines.

In a recent overview of the 'greed' versus 'grievance' debate, Murshed and Tadjoeddin (2009: 96-99) identify, alongside Stewart and Østby, Ted Gurr (1970, 2000) as one of the most influential 'grievance' academics. Gurr stresses that group identity is important in creating shared preferences, which in turn allows groups to adopt certain rational strategies. Through the forging of group identity, Gurr attempts to explain rational choice of political violence for different ethnic and/or religious groups. He identifies "four main determinants of civil war: (a) the salience of ethnocultural identity, as it relates to other types of socio-economic identities; (b) the level of grievance (actual or expected); (c) the capacity of ethnopolitical groups to mobilize (a function of their cohesion); and (d) the available opportunities for political action by each group"([Gurr] in Sambanis 2002: 224).

Murshed and Tadjoeddin (2009: 96-99) identify a number of other scholars writing from a 'grievance' perspective, however, this paper will focus on the above mentioned scholars for want of space and relevance.

Despite the fact that the above mentioned theories largely focus on civil war onset, research suggests that the causal mechanisms underlying both onset and duration are largely similar (Elbadawi & Sambanis 2002: 307-308, Bleany & Dimico 2009: 29).

Returning to Muggah's conflict adapted IRLR model, we can single out a number of components which are expected to relate most to grievance theories of civil war in relation to the Colombian context; landlessness, economic marginalisation, declines in political participation and political violence. As Elhawary (2007: 4-6) notes, from a historical perspective disputes over land seem to be the most continual and largest source of grievance. Land owned by subsistence farmer peasants has traditionally been colonized by landowning elites. The peasants that previously owned this land were either forced to become wage labourers or flee. Elhawary notes that the type of conflict differs by region. Thus, in regions where property rights are properly defined, conflict is concerned with the height of wages and working conditions. In regions, where property rights are not well defined, conflict tends to revolve around ownership. Guerrillas were originally employed by the peasantry as a form of self defence against this land owning elite and also drug traffickers. They established themselves particularly in areas where land disputes existed. Paramilitaries on the other hand were created to protect the landowning elite from the leftist guerrillas. Particularly in the case of paramilitaries, social conflict started being increasingly replaced by the struggle for territorial dominion and was often accompanied by the expulsion of large amounts of peasants. This process has exacerbated land inequality to such a degree that Colombia had a Gini coefficient of 0.85 for land inequality in 2004. Stewart (2010), deems land inequality in agricultural societies, to be of particular importance in fuelling grievances prior to, or during, a civil war.

Economic marginalisation, is greatly related to the land colonisation issue. It is estimated that paramilitaries own over 50% of Colombia's most fertile and resource rich land, often having displaced the original inhabitants. (Elhawary 2007: 5). This has led to gross inequalities and impoverishment in the countryside, where approximately 85 percent of the inhabitants live beneath the poverty line (Dion & Russler 2008: 408). According to Minear (2006: 1): "pervasive poverty and chronic underdevelopment underlie conflict-related deprivation and depredation." However, economic marginalisation does not appear to be a strong determinant of whether poor people will join either left-wing guerrillas or right-wing paramilitaries. In fact, people perceiving themselves as poor are more likely to join counterinsurgents than rebel groups (Arjona & Kalyvas 2009: 29).

Political violence has been mentioned with regards to the assassination of numerous Unión Patriótica party members in the 1980's. However it is not only politicians that have been targeted. Similarly, community leaders, human rights defenders, small business persons, trade unionists, religious figures and even ordinary citizens have also been on the receiving end of politically oriented violence and kidnappings by either paramilitaries or guerrillas (Minear 2006: 19, HRW 2010: 45). Political assassinations has led to an extremely insecure environment regarding political participation, something Walter (2004: 372), from a grievance perspective argues, has a significant effect on the likelihood of continuity, or renewal of civil war.

A major incongruity, however, when applying the grievance model of civil war to the Colombian conflict, is the non-ethnic nature of the war. Thus, grievances in this context are vertical rather than horizontal. Nonetheless, Afro-Colombians and Indigenous communities are particularly marginalised (Wouters 2001, Hristov 2009), in addition most large landowners are disproportionately from Spanish descent. Mixed-race (mestizo) peasants that have traditionally worked on the farms of the landowning elite, enjoy virtually no social mobility (Williamson 1965). People from these three marginalised groups are the most likely to be displaced (Albuja 2010: 3). Although Colombia's conflict does not run along ethnic lines, a recent CRISE[4] finding, shows that recruits were most likely to join either paramilitary or rebel organisations, depending on which controlled the area the recruit lived in (Arjona & Kalyvas 2009: 30-31). This raises questions about the definition of ethnicity. It has been extensively argued that ethnicity is largely based on the social construct of identity, depending on the interpretation of communities and manipulation by its leaders (Hopf 1998: 176, 182, Stewart 2010: 7). The most famous example of this is of the Lue 'ethnic' community in Thailand who were similar to their neighbours in every way (language, trait distributions, etc.), yet considered themselves to be a distinct ethnic community, solely on the basis of their own interpretation (Moerman 1965). From this point of view, one could argue that a certain

---

[4] Centre for Research on Inequality, Human Security and Ethnicity

group identity or association is created with both paramilitaries and guerrilla groups, in the local areas in which they operate. Thus, as one interviewee in the CRISE report states: "I saw [FARC combatants] since I was little. I liked what they used to do." Whilst another says: "since I was a kid I knew the FARC, and I liked weapons." (Arjona & Kalyvas 2009: 31). This is not a new trend, as this paper has mentioned previously: as far back as the 19th century peasants chose to support either Conservative or Liberal factions depending on who controlled the area they lived in. Arjona (2005) gives a good overview of rebel and counterinsurgent recruitment in such 'micro-orders'.

In light of these social constructivists and anthropological arguments, the Colombian conflict might fit grievance theories of civil war to a greater extent than first appears. Seemingly vertical inequalities get a distinctly more horizontal slant when ethnicity is replaced by identity. Gurr's 'four determinants of civil war' suddenly provide a better fit: (a) the salience of ethno-political groups as they relate to other socio-economic identities is replaced by the salience of 'micro-orders' and how they provide communities with an identity, (b) the level of grievance in rural Colombia has been discussed above, (c) the capacity for ethnopolitical groups to mobilise has also been discussed previously: peasant groups effectively formed into guerrilla movements, whilst land owners were able to mobilise peasants into paramilitary forces (not along ethnic lines, but because peasants were part of a 'micro-order'), (d) the available opportunities for political action by each group (especially that of leftists) have become seriously undermined, due to political intimidation and assassination.

Grievance theories of civil war, as presented by Stewart and Gurr, do not appear to fit the Colombian context well. If however one replaces ethnicity with group identity, the different independent variables put forward by both authors suddenly become relevant. With regards to the conflict adapted IRLR model, we can identify four significant areas correlating to those variables; issues pertaining to landlessness, economic marginalisation, declines in political participation and political violence. These can be related back to two significant horizontal inequalities discussed by Stewart (2010): economic and political inequality, as well as determinants (b) and (d) in relation to Gurr (2000).

Greed Theories of Civil War

Greed based theories of civil war onset and duration have their roots in micro-economic theories of individual rational decision making. The main difference with grievance based theories, is that they argue from an opportunity or feasibility point of view. Civil war is not driven by a motivation to redress earlier grievances, but will occur if specific variables are conducive to permitting rebellion. The most notable contemporary academics arguing from this point of view are Paul Collier and Anke Hoeffler (1998, 2000, 2004) and Collier et al. (2004, 2009). Other prominent academics include James Fearon (2004) and his earlier work with David Laitin (2003), Jack Hirshleifer (1995) and Michael Ross (2003, 2004).[5] The Collier and Hoeffler model is the most prominent within donor circles and the Western media (Murshed & Tadjoeddin 2009: 88). Their model has also been widely consulted with regards to disaster risk reduction of complex emergencies (Alexander 2006: 4). For these reasons, their scholarly contributions will be the main focus of this section of the paper.

The studies mentioned above are quantitative in nature and therefore would appear not to be suitable for single case studies as in this paper. However, they can serve an illustrative purpose with regards to how certain independent variables influence the dependent variable, which in this case is civil war onset. This method has been used by academics such as Stefan Blum (2006), who illustrates the effectiveness of Collier and Hoeffler's 2004 model, by calculating the contributions to risk of civil war onset of all the individual independent variables put forward in the core regression model. He finally arrives at two overall percentages with regards to probability of civil war onset for the countries of Uganda and Tanzania. This paper will take a similar approach in order to illustrate the importance of the cocaine trade, which in the Collier and Hoeffler model *would be* proxied for by primary commodity exports (PCE) as a function of GDP (as discussed below). Although Collier and Hoeffler's 2004 model is only applicable to civil war onset, Collier et al. (2004) findings on civil war duration, similarly place great weight on the importance of lootable resources for rebel financing in the form of PCE.

As mentioned in the introduction, Gleditsch' (2004) revision of the Correlates of War (COW) project (used in Collier and Hoeffler's latest model) denotes a period of non-conflict between 1993 and 1998 for Colombia. This paper looks at the predictive power of the Collier and Hoeffler's 2004 model with regards to civil war onset in 1998. These results are then applied to Muggah's IRLR model, which will identify specific areas of focus for both future

---

[5] See Murshed & Tadjoeddin (2009: 89-96) and Hegre (2004) for an overview of the debate.

DRR as well as current advocacy and assistance strategies. These results will also be useful with regards to civil war duration variables, as Collier et al. (2004: 264) note: "Although the concept of conflict continuation is not the same as that of conflict re-ignition, it is reasonable to suppose that the risk of conflict re-ignition, which is latent during conflict, broadly follows the observed risk of conflict continuation." In their paper on civil war duration, Collier et al. similarly find PCE to be of significant influence.

Collier and Hoeffler see participation in rebellion as rational decision making dependent on the demand of the so called 'rebel market'. Civil wars occur because of three interacting factors, namely: preference, opportunity and perception. So called rebel entrepreneurs make a cost-benefit analysis based on the perception of opportunity and the preference of private gain. The different independent variables put forward by Collier and Hoeffler's 2004 model are discussed here:

One of the consistently recurring corner stones of the Collier and Hoeffler model is the availability of "lootable resources" which, as mentioned above, is measures through primary commodity exports as a percentage of GDP. Their model finds a quadratic relationship between PCE and civil war onset, peaking at 33%. The assumed causal mechanism behind this relationship, is that primary commodities are an easily lootable resource with which to finance rebels. In addition, countries with high levels of natural resources are often rent-seekers and thus do not need to tax their citizens heavily, which consequently makes them less accountable, which can increase grievances. The relationship is quadratic, because an over abundance of rents often leads to the state having a very powerful coercive apparatus or gives them the ability to co-opt opposition groups. Collier and Hoeffler also find a statistically significant relationship between civil war onset and GDP growth. This presents itself in the form of a negative relationship, since an increase in GDP means a tighter labour market and thus less opportunity for a rebel organisation to recruit. Another independent variable; the percentage of male secondary school enrolment, is a similar proxy to GDP growth, as they argue school-goers will forgo investment and future earnings by joining a rebellion. Collier and Hoeffler also argue that mentality changes brought on by education can withhold youths from joining civil conflict.

Other, less relevant (to this paper) and less significant variables that have survived the stepwise deletion to form part of the core model, are; time since previous civil war, geographic dispersion of the population, social fractionalisation, ethnic dominance and population size. Contrary to grievance theories of civil war, the greed model does not find a strong correlation between civil war onset and ethno-linguistic and religious fractionalisation.

11

In fact, they argue that ethnic dominance might be a stronger indicator of civil war risk, yet still relatively insignificant compared to the availability of lootable resources.

The risk of civil war onset using the Collier Hoeffler model is calculated by employing the following method: independent variables, taken from the Collier and Hoeffler model, are multiplied by their respective coefficient, producing the independent variable's contribution to risk. The sum of these are then entered into the formula of appendix A2.3 in Collier and Hoeffler's 2004 work:

$$\hat{p}_{it} = \frac{e^{\hat{W}_{it}}}{(1 + e^{\hat{W}_{it}})} \cdot 100$$

$\hat{p}_{it}$ representing the risk of civil war onset in percent.

The following table[6] gives the percentage of the probability of civil war onset for Colombia in 1998 as well as a similar calculation for the 1980 civil war in Uganda taken from Blum (2006) for comparative reasons:

Table 1				Contribution to Risk	
	Colombia 1998	Uganda 1980	Coefficient	Colombia	Uganda
Primary commodity exports/GDP	0.039955	0.273	18.937	0.75662	5.169801
Primary commodity exports/GDP squared	0.001596	0.0745	-29.443	-0.047	-2.19436
Male secondary school enrolment (% of total in school)	66	7	-0.032	-2.112	-0.224
Annual GDP growth during previous 5 years	1.8	-2.922	-0.115	-0.207	0.33603
Months of peace since previous civil war	60	162	-0.004	-0.24	-0.648
Geographic dispersion of population index	0.647	0.508	-2.487	-1.60909	-1.2634
Social fractionalization index	960	5,940	-0.0002	-0.192	-1.2634
Ethnic dominance	1	0	0.67	0.67	0
Ln Population	37,782,390	12,806,841	0.768	13.39957	12.5687
Constant	1	1	-13.073	-13.073	-13.073
Total				-2.6539	-0.51623
Probability of civil war onset (%)				6.6	37.4

---

[6] Sources used for Tables 1 and 2 are listed in appendix 3: Collier & Hoeffler (2004:592-595)

As the comparative table above shows, the Collier and Hoeffler model does not appear to give a very good fit with regards to the conflict in Colombia. One could argue however, that this is primarily due to the obvious fact that the World Bank does not take the export of illicit narcotics into account when measuring PCE. In addition, should illicit drugs be taken into account in the model, the PCE relationship would probably not be quadratic, as it is highly unlikely that the government has *directly* benefitted from the drug trade. The importance of Colombia's drug trade in relation to the civil conflict has frequently been argued by academics (LeoGrand & Sharp 2000, Klare 2001, Ross 2004, Minear 2006). Collier and Hoeffler (2004: 565) argue - despite their model not factoring in drug exports - that narcotics do play a significant role with regard to rebel financing and incentives. In addition, the Colombian and especially the US government are convinced that the drug trade and 'terrorism' are inextricably linked, and have adjusted their policies to fight terrorism in Colombia accordingly (Minear 2006: 10). The results of the Collier and Hoeffler model after factoring in the impact of the drug trade, appear to disagree with this argument. Roberto Steiner (1998) estimates that in 1998, illegal drugs comprised roughly 3% of GDP and 25% of Colombian exports. With these figures PCE/GDP for 1997 (assuming the figures were relatively similar to 1998) becomes 8.6%, and when inserted into the model produces the following outcomes:

Table 2			Contribution to Risk
	Colombia 1997	Coefficient	Colombia
Primary commodity exports/GDP	0.086738	18.937	1.642548
Primary commodity exports/GDP squared	0.007523	-29.443	-0.22151
Male secondary school enrolment (% of total in school)	66	-0.032	-2.112
Annual GDP growth during previous 5 years	1.8	-0.115	-0.207
Months of peace since previous civil war	60	-0.004	-0.24
Geographic dispersion of population index	0.647	-2.487	-1.60909
Social fractionalization index	960	-0.0002	-0.192
Ethnic dominance	1	0.67	0.67
Ln Population	37782390	0.768	13.39957
Constant	1	-13.073	-13.073
Total			-1.94248
Probability of civil war onset (%)			12.5

Although the probability of civil war onset in the case of Colombia doubles to 12.5% when adjusted for illicit drug export, it is still not as substantial as Uganda in 1980, which reaches 37.4%. Even if we were to view illicit drugs, proxied by PCE, as a linear function by negating the negative contribution from PCE/GDP squared (as it is not directly supplying the government with revenue), the probability of civil war onset would only rise to 15.2%. According to the Collier & Hoeffler model however, a country is not considered to be at risk until it is predicted to have a probability of around 20 percent or more (Zinn 2005: 89).

The Collier and Hoeffler model, in addition to not providing a satisfactory fit with regards to civil war onset in Colombia, only contains two variables which are useful in relation to the IRLR model, namely: PCE/GDP and males secondary school enrolment. If humanitarian organisations would like to have a maximum impact on stopping the current, or preventing future, civil war in Colombia, they should focus on areas of economic marginalisation and derivatives thereof such as joblessness and food insecurity, in addition to access to education. Addressing economic marginalisation of the countryside by encouraging farmers to grow crops other than coca or poppies is a far more effective way of tackling Colombia's drug problem than aerial spraying, according to the United Nations Office on Drugs and Crime (UNODC 2010). By persuading farmers to grow alternative crops, the economic incentive for people to join illegal armed parties is reduced, at least according to the rationale offered by Collier and Hoeffler. Similarly, higher school enrolment would have a similar effect; if Colombia would have had a 100% ratio of male secondary school enrolment in 1997, it would have reduced the risk of civil war onset by 4.3%.

With regards to the causal mechanisms put forward by Collier and Hoeffler (2004), the results appear ambiguous. There is some evidence that "lack of reference points such as family life and schools, young people … are gradually swallowed up by youth gangs and subsequently paramilitary groups." (IDMC 2006: 61). This is partly compatible with the above given causal mechanism for the proxy of male secondary school enrolment as argued by Collier and Hoeffler. With regards to individual economic incentives for entering civil conflict the result is similarly inconclusive. The previously mentioned CRISE study conducted by Arjona and Kalyvas (2009: 29), finds that 50% the ex-paramilitary and 25% of ex-FARC combatants mentioned economic incentives in response to the open ended question: 'What was most important in your decision to enlist?'

Conclusion

This paper set out to identify the most significant contributing factors to the phenomenon of civil war in Colombia, with the goal of implementing those findings into strategies for relief provision, as well as informing strategies for disaster risk reduction and advocacy. In the introduction, the following two hypotheses were proposed:

H1:   *Greed theories of civil war better explain the Colombian conflict*

H2:   *Goal-oriented humanitarian organisations should focus on those components found in Muggah's (2000) adapted IRLR model, which correspond to independent variables identified in greed theories of civil war relevant to the Colombian context, in order to effectively implement strategies of relief provision, in addition to DRR and advocating*

Based on the findings of this paper, both hypotheses have been proved wrong.

Although Colombia's historical context contains variables which are put forward by both 'greed' and 'grievance' scholars as significant in relation to civil war, neither tradition can be said to fully fit the current conflict. Grievance theories of civil war can identify a number of variables, which they argue fuels the conflict in Colombia; land disputes resulting in landlessness, economic marginalisation and declines in political participation largely due to political violence. However, the major incongruence with grievance theories is that these issues of deprivation are not relative or horizontal. In other words, they do not run along ethnic lines. However, if we were to look at ethnicity as a form of identity and apply this broader concept to the Colombian case, we would get a good fit. Examining the conflict in Colombia at the micro-level, specifically with regards to recruitment, we find that people decide to join either counterinsurgents or rebels, depending on who controls the area (so called 'micro-orders'). This also conforms to a long standing pattern throughout Colombian history.

Greed theories on the other hand, provide neither a good fit, nor do their causal mechanism for the relevant variables prove correct. Collier and Hoeffler's 2004 model has been widely used in policy circles, such as the World Bank and national governments and is therefore very relevant to this paper. However, when applied to the context of Colombia, the results are not convincing. Even when the export of illicit drugs is factored in to the 'primary commodity/GDP' variable *and* the negative effects of government rent-extraction are ignored, the result remains at 15.2%, which is 4.8% lower than countries which are deemed to be 'at risk'. Collier and Hoeffler's proposed causal mechanisms of independent variables which correspond to components of the IRLR model used in this paper - economic marginalisation and access to education - are also unsatisfactory. Although there is some evidence that non-school goers are more likely to join armed movement, the assumption that most people join for private material gain is untrue, especially in the case of guerrillas.

In conclusion, this paper suggests that goal-oriented humanitarian organisations focus on the following components of Muggah's (2000) conflict adjusted IRLR model, in order to have a maximum impact on the root causes of Colombia's civil war, mainly with regards to implementing strategies of relief provision, but also in relation to strategies of DRR and advocacy: landlessness, economic marginalisation, declines in political participation and political violence. It must also be noted that a similar focus on access to education could reduce the severity of the war, whilst areal spraying merely exacerbates the problem. Future research on this subject could prove fruitful with regards to the relationship between civil war, 'micro-orders' and the current Colombian conflict.

Reference List

Albuja, Sebastián. (2010) 'Towards property restitution for IDPs in Colombia', *Internal Displacement Monitoring Centrum*. Building Momentum for Land Restauration

Alexander, David. (2006) 'Globalization of Disaster: trends, problems and dilemmas', *Journal of International Affairs* 59(2): pp. 1-22

Arjona, Ana. (2005) 'Understanding Recruitment in Civil War', *Yale University*. Paper presented at the Workshop of the Program on Order, Conflict and Violence.

Arjona, Ana and Kalyvas, Stathis. (2009) 'Rebelling Against Rebellion: Comparing Insurgent and Counterinsurgent Recruitment', *Centre for Research on Inequality, Human Security and Ethnicity*. CRISE Workshop: Mobilisation for Political Violence: What do We Know?

Barnett, Michael. (2005) 'Humanitarianism Transformed', *Perspectives on Politics* 3(4): pp. 723-740

BBC. (2010a) 'Colombia Timeline', http://news.bbc.co.uk/2/hi/americas/1212827.stm [last accessed 12/11/2010]

BBC. (2010b) 'Colombia: Country Profile', http://news.bbc.co.uk/2/hi/americas/country_profiles/1212798.stm [last accessed 12/11/2010]

Bleaney, Michael and Dimico, Arcangelo. (2009) 'Incidence, Onset and Duration of Civil Wars: A Review of the Evidence', *Centre for Research in Economic Development and International Trade*. Research Paper

Blum, Stefan. (2006) 'East Africa: Cycles of Violence, and the Paradox of Peace', *ZEF Discussion Papers on Development Policy*. 107

Carillo, Angela. (2009) 'Internal Displacement in Colombia: Humanitarian, Economic and Social Consequences in Urban Settings and Current Challenges', *International Review of the Red Cross* 91(875): pp. 527-546

Cernea, Michael. (1997) 'The Risks and Reconstruction Model for Resettling Displaced Populations', *World Development* 25(10): pp. 1569-1588

Collier, Paul. (2000) 'Rebellion as a quasi-criminal activity', *Journal of Conflict Resolution*. 44(6): pp. 839-853

Collier, Paul and Hoeffler, Anke. (1998) 'On economic causes of civil war', *Oxford Economic Papers*. Vol. 50: pp. 563-573

Collier, Paul and Hoeffler, Anke. (2000) 'Greed and Grievance in Civil War', *World Bank Policy Research Paper*. 2355

Collier, Paul and Hoeffler, Anke. (2004) 'Greed and Grievance in Civil War', *Oxford Economic Papers* 56(4): pp. 563-595

Collier, Paul, Hoeffler, Anke. and Rohner, Dominic. (2009) 'Beyond Greed and Grievance: feasibility and civil war', *Oxford Economic Papers* 61(1): pp. 1-27

Collier, Paul., Hoeffler, Anke. and Söderbom, Måns. (2004) 'On the Duration of Civil War', *Journal of Peace Research* 41(3): pp. 253-273

Deng, Francis. (1998) 'Guiding Principles on Internal Displacement submitted by Francis Deng, Special Representative of the Secretary-General to the UN Commission on Human Rights', *International Journal Refugee Law* 10(3): pp. 563-572

Dion, Michelle and Russler, Catherine. (2008) 'Eradication Efforts, the State, Displacement and Poverty: Explaining Coca Cultivation in Colombia during Plan Colombia', *Journal of Latin American Studies* 40(3): pp. 399-421

Elbadawi, Ibrahim and Sambanis, Nicholas. (2002) 'How Much War Will we see?: Explaining the Prevalence of Civil War', *Journal of Conflict Resolution* 46(3): pp. 307-334

Elhawary, Samir. (2007) 'Between War and Peace: land and humanitarian action in Colombia', *HPG Working Paper*. Overseas Development Institute

Fearon, James. (2004) 'Why Do Some Civil Wars Last So Much Longer Than Others?', *Journal of Peace Research* 41(3): pp. 275–301

Fearon, James and Laitin, David. (2003) 'Ethnicity, Insurgency, and Civil War', *American Political Science Review* 97(1): pp. 75–90

Fox, Fiona. (2001) 'New Humanitarianism: Does It Provide a Moral Banner for the 21st Century?', *Disasters* 25(4): pp. 275-289

Gleditsch, Kristian. (2004) 'A revised list of wars between and within independent states, 1816–2002', *International Interactions* 30(3): pp. 231–262

Gurr, Tedd. (1970) *Why Men Rebel*. (Princeton University Press: Princeton)

Gurr, Tedd. (2000) *Peoples Versus States: Minorities at Risk in the New Century*. (US Institute of Peace: Washington DC)

Hegre, Håvard. (2004) 'The Duration and Termination of Civil War', *Journal of Research Peace* 41(3): pp. 243-252

Hirshleifer, James. (1995) 'Anarchy and its breakdown', *Journal of Political Economy* 103(1): pp. 26–52.

Hopf, Ted (1998) 'The Promise of Constructivism in International Relations Theory', *International Security* 23(1): pp. 171-200

Hristov, Jasmin. (2009) 'Social Class and Ethnicity/Race in the Dynamics of Indigenous Peasant Movements: the Case of the CRIC in Colombia', *Latin American Perspectives* 36(4): pp. 41-63

HRW. (2010) 'Paramilitaries' Heirs: the new face of violence in Colombia', http://www.hrw.org/node/88060 [last accessed30/10/2010]

Ibáñez, Ana María and Vélez, Carlos. (2008) 'Civil conflict and forced migration: the micro determinants and welfare losses of displacement in Colombia', *World Development* 36(4): pp. 659–676

IDMC. (2010) 'Government "Peace Process" Cements Injustice for IDPs', http://www.unhcr.org/refworld/publisher,IDMC,,COL,44bf3b4d4,0.html [last accessed 23/11/10]

Klare, Michael. (2001) *Natural Resource Wars: The New Landscape of Global Conflict* (Metropolitan Books: New York)

Leech, Garry. (2002) *Killing Peace: Colombia's Conflict and the Failure of U.S. Intervention*. (New York: Information Network of the Americas)

LeoGrande, William and Sharpe, Kenneth. (2000) 'Two Wars or One?: Drugs, Guerrillas and Colombia's New *Violencia*', *World Policy Journal* 17(3): pp. 1-11

Minear, Larry. (2006) 'Humanitarian Agenda 2015 Colombian Country Study', *Briefing Paper*. Feinstein International Center

Moerman, Michael. (1965) 'Ethnic Identification in a Complex Civilization: Who Are the Lue?', *American Anthropologist* 67(5): pp. 1215-1230

Muggah, H.C.R. (2000) 'Conflict-induced Displacement and Involuntary Resettlement in Colombia: Putting Cernea's IRLR Model to the Test', *Disasters* 24(3): pp. 198-216

Murshed, Syed, and Tadjoeddin, Mohammad. (2009) 'Revisiting the greed and grievance explanations for violent internal conflict', *Journal of International Development* 21(1): pp. 87-111

Østby, Gudrun. (2008) 'Polarization, Horizontal Inequalities and Violent Civil Conflict', *Journal of Peace Research* 45(143): pp. 143-162

Rogers, Daniel. (2002) 'Colombia and the United States: Providing for their "Common" Defence', *The Nature of War Seminar*. National War College

Ross Michael. (2003) 'Oil, drugs and diamonds: the varying role of natural resources in civil wars', in *The Political Economy of Armed Conflict: Beyond Greed and Grievance*, Karen B, Sherman J (eds). Lynne Rienner (Boulder, CO): pp. 47–70

Ross Michael. (2004) 'What do we know about natural resources and civil wars', *Journal of Peace Research* 41(3): pp. 337–356

Sambanis, Nicholas. (2002) 'A Review of Recent Advances and Future Directions in the Quantitative Literature on Civil War', *Defence and Peace Economics*.13(3): pp. 215-243

Steiner, Roberto. (1998) 'Colombia's Income from the Drug Trade', *World Development* 26(6): pp. 1013-1031

Stewart, Frances. (2002) 'Horizontal Inequalities: A Neglected Dimension of Development', *Queen Elizabeth House, University of Oxford.* Working Paper Number 81

Stewart, Frances. (2010) 'Horizontal inequalities as a cause of conflict: a review of CRISE findings', *Centre for Research on Inequality, Human Security and Ethnicity.* Number 1, January 2010

UNHCR. (2010) 'UNHCR Eligibility Guidelines for Assessing the International Protection Needs of Asylum-Seekers from Colombia', http://www.unhcr.org/refworld/docid/4bfe3d712.html, [last accessed 24/11/2010]

UNODC. (2010) 'Desarollo Alternativo en el Área Andina', http://www.unodc.org/documents/alternative-development/Desarollo_alternativo.pdf [last accessed 23/11/2010]

Uvin, Peter. (1999) 'The Influence of Aid in Situations of Violent Conflict', http://www.ndu.edu/itea/storage/610/Impact%20of%20Aid%20Uvin.pdf, [last accessed 23/11/2010].

Walter, Barbara. (2004) 'Does Conflict Beget Conflict? Explaining Recurring Civil War', *Journal of Peace Research* 41(3): pp. 371-388

Williamson, Robert (1965) 'Toward a Theory of Political Violence: The Case of Rural Colombia', *The Western Political Quarterly* 18(1): pp. 35-44

Wouters, Mieke. (2001) 'Ethnic Rights Under Threat: The Black Peasant Movement Against Armed Groups' Pressure in the Chocó, Colombia', *Bulletin of Latin American Research* 20(4): pp. 498-519

Zinn, Annalisa. (2005) 'Theory versus Reality: Civil War Onset and Avoidance in Nigeria Since 1960', in *Understanding Civil War*, Collier, Paul and Sambanis, Nicholas (eds). (World Bank: Washington DC)